Elaine Rhodes

MW00956125

ASIC
BASICS

An Introduction to Developing
Application Specific Integrated Circuits

Black & white edition

— Lulu Press —

For a full-color edition of this guide, visit
http://www.lulu.com/content/118763

First Lulu edition April 11, 2005
Black & white edition May 1, 2005
Published by "Published by Lulu" May 27, 2008

ISBN 978-1-4357-1910-1

This book was produced on a Windows PC using Adobe Framemaker.
Illustrations and photographs were prepared using Adobe Illustrator and Adobe Photoshop.
Fonts used are Goudy Old Style, Lucinda Sans, and Curlz MT.
Georgia is used on this page and the back cover.

Contents

Introduction

Did you ever wonder what magic engineers use to make cell phones, music players, and a plethora of other products smaller and less expensive every year, even while adding amazing new features and capabilities? They use the magic of integrated circuits (ICs), those tiny silicon chips that contain thousands of electronic circuits. Every year, engineers pack more circuits into ICs by inventing new ways to advance the optical imaging and etching technologies they use create microscopic devices on the surfaces of silicon chips. Engineers at Intel packed their latest Pentium chip with over one hundred million transistors!

Companies like Intel dedicate teams of over a hundred engineers to develop complex ICs like Pentium processors. Only a handful of giant corporations around the world can afford to invest so heavily to develop ICs. These companies see their investments pay off handsomely because they sell millions of chips annually, generating enough revenue to pay back the development expenses and earn a good profit. How can thousands of other companies gain the benefits of IC technology when their products do not generate such large revenues? They need some way to develop ICs at a dramatically lower cost.

IC manufacturers satisfy the need for lower cost IC development with a technology called application specific integrated circuits, or ASICs (pronounced "Ay-six"). ASICs lower the cost of developing an IC for a specific application by sharing a standard basic design among many applications. An engineer develops an ASIC by building on top of a standard base chip, much like a pizza is customized with pepperoni, mushrooms, and other toppings. The ASIC engineer's job is much less work than developing an entirely customized or *full custom* IC, just like picking the toppings for a pizza is easier than making the whole pie from scratch. But because ASIC engineers rely on standard base designs, they cannot pack the circuits as efficiently as on a full custom IC. Therefore, an ASIC part is larger and costs more than an equivalent full custom IC. Nevertheless, ASICs deliver most of the benefits of IC technology at a reasonable price, while being developed much more quickly and cheaply than full custom ICs.

With ASIC technology, a single engineer can develop a complex IC with hundreds of thousands of transistors in a reasonable amount of time, usually about a year, at a cost and quality level suitable for mass production. ASICs bring the magic of ICs to products like medical equipment, tape drives, professional video recorders, and satellites — just about any product shipping less than a million units annually.

If you are an engineer embarking on your first ASIC project, or if you just curious about ASICs, keep reading to learn what ASICs are and how they are developed, a process most engineers find challenging, fun, and rewarding.

This guide is divided into the following sections:

- **Getting Started.** Explains what ASICs are, how they are manufactured, and what role you, as an engineer, play in developing them.
- **Developing an ASIC.** Describes the process flow you follow to develop an ASIC, including instructions for the procedures you execute.
- **Moving On.** Suggests how you can learn more about developing ASICs.
- **Troubleshooting.** Describes problems you are likely to encounter as you develop ASICs, and suggests solutions.

So, grab a tasty piece of pizza and dive in!

Appetizer

Getting Started

You already have an idea of what ASICs are, but before you start developing your first ASIC, you should know the answers to these questions, which are discussed in this section:

- **How are ASICs made?** Explains how ASICs and other ICs are manufactured so you can answer the next question.

- **Are ASICs right for my project?** Describes several classes of ICs and explains how you choose the best class to meet your needs.

- **What does an ASIC engineer do?** Describes the engineer's responsibilities to design logic expressed in high-level description languages or HDLs, and how they turn their HDL code into silicon chips.

- **How do I choose an ASIC vendor?** Explains the criteria you use to select a vendor to provide your ASICs.

- **What tools and equipment do I need?** Describes the CAD (computer aided design software) tools you use to develop ASICs.

How are ASICs made?

ASICs and other ICs are made the same way, which is described below. What distinguishes ASICs from other types of ICs is which parts of the manufacturing process are customized for a particular design and which parts are standardized across many designs. To understand this concept better, you need to know how ICs are manufactured.

Fabricating an IC is like making a pizza. To make a pizza, you start with a crust, add a layer of sauce, a layer of cheese, and finally toppings like pepperoni and green peppers. An IC is made in a similar fashion. The IC's "crust" is a round, flat silicon wafer. The wafers used in the latest IC fabrication facilities or fabs (IC manufacturing plants) are 300 millimeters or 12 inches in diameter, the size of a small pizza. On top of this wafer, circuits are built up in layers of materials like polysilicon, aluminum, and silicon dioxide, as shown below:

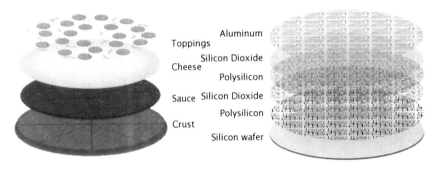

Building pizzas and ICs in layers

An important difference between making a pizza and making an IC is a process called etching. While a layer of cheese covering the entire pizza is great, a layer of material covering the entire surface of a silicon wafer is not very useful. Patterns must be etched into the material to create transistors and interconnections (wires) on the surface of the wafer.

Etching creates microscopic patterns on the wafer's surface, and is the true magic of IC technology. Etching delineates fine geometries using an optical process, forming features ten thousand times narrower than the period at the end of this sentence.

The etching process is performed with the following steps:

1. The etching process starts with a new silicon wafer, or one that already has some etched layers.

————— Silicon wafer

2. A layer of material is deposited over the entire surface of the wafer.

————— Material to be etched

3. The layer of material is coated with a photoresistive chemical or *photoresist*.

————— Photoresist

4. Light is focused on the photoresist through a mask, which is like a photographic negative of the patterns to be etched into the material.

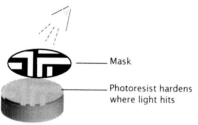
————— Mask

————— Photoresist hardens
where light hits

5. Because of its photoresistive properties, the photoresist hardens where the light hits it, as shown above.

6. The photoresist that was not hardened is washed away leaving photoresist in the desired pattern on the surface of the material.

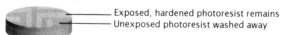
————— Exposed, hardened photoresist remains
————— Unexposed photoresist washed away

7. The material is chemically etched using the photoresist as a stencil, forming the mask's pattern in the material.

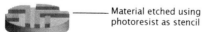
————— Material etched using
photoresist as stencil

Once the desired patterns have been etched into the material, the layer is complete. The etching process is repeated with a new material and a new mask to make the next layer. Advanced IC manufacturing processes use over twenty masks to make a chip. That's a lot of toppings on your pizza!

The last step in making a pizza is to slice it and put it in a box. The IC wafer must be sliced up, too, into individual *die*, or little rectangular chips. (*Die* is both singular and plural.) Each die is one IC, so a wafer yields more ICs if the die is small. ICs with smaller die sizes cost less than those with larger die sizes because processing a wafer costs the same amount regardless of how many die are on it.

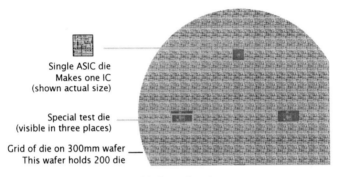

Single ASIC die
Makes one IC
(shown actual size)

Special test die
(visible in three places)

Grid of die on 300mm wafer
This wafer holds 200 die

IC die and wafer

The die are tested while the wafer is still intact. A robotic machine performs the tests rapidly and leaves a drop of red ink on any die that fails the test. The wafer is *diced* with a diamond-tipped circular saw or a hydraulic knife, which is a thin, high pressure stream of water. The die with red ink dots are discarded and the rest are put into appropriate chip packages, producing the familiar little black plastic or ceramic "bugs" you see on electronic circuit boards. Examples of IC packages are pictured below:

14-pin DIP "bug"
Dual in-line package

44-pin PLCC
Plastic leadless chip carrier

100-pin QFP
Quad flat pack

144-pin BGA
Ball grid array

Some IC packages (shown actual size)

Are ASICs right for my project?

Now that you know how ICs are manufactured, you can decide whether ASICs are the best kind of ICs to use for your project. ASICs are one of three classes of ICs that manufacturers have invented to allow you to optimize for:

- The cost of the parts themselves.
- The NRE you have to pay. (NRE stands for non-recurring engineering expense, the fixed cost of developing the IC.)
- The amount of time it takes to develop the IC.

The three IC classes, full custom, ASIC, and FPGA, are distinguished by how much of the manufacturing process is customized for your particular project, and how much is standardized for many designs. These classes of ICs are described below.

Full custom ICs

Full custom ICs are like pizzas made to order, where you pick the crust, the sauce, the cheeses, and the toppings. With full custom ICs, engineers customize all the mask layers.

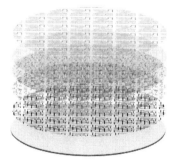

Full custom: All layers customized

The objective of developing full custom ICs is to achieve a low part cost. The engineering team uses creativity and as much time as they need to pack the die as tightly as possible. Engineers design twenty or more masks, so the job takes a long time and incurs a large NRE. The result is a small die size, so the cost of manufacturing each part is low. Small full custom chips may cost twenty-five cents or less, while large chips like Pentium processors cost tens or hundreds of dollars.

ASICs

ASICs are like pizzas where the bases are pre-made and all you pick are the toppings. With ASICs, engineers customize only the last few mask layers. These layers interconnect elements ASIC manufacturers predefine in standard base wafers.

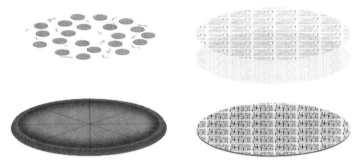

ASIC: Standard base, only "toppings" customized

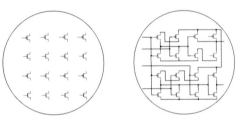

Transistors on base wafer   Transistors wired into a circuit
completing the ASIC

Customizing an ASIC, schematic view
An abstract representation of ASIC customization

The objective of developing ASICs is to reap the benefits of IC technology without investing the time and expense required for full custom chips. ASICs can be developed more quickly and with a lower NRE than full custom chips because engineers only design a few masks. However, the ASIC base wafer has a fixed set of configurable elements, so ASIC engineers do not have the flexibility to pack the die as tightly as possible. Therefore, a design in an ASIC has a larger die size and higher part cost than the same design in a full custom IC. ASICs typically cost fifty cents to two dollars in mass production, but very large ASICs can have over a million gates and cost hundreds of dollars. An ASIC costing $900 may seem very expensive, but if it is used in a system that sells for $50,000, it can still be cost effective.

FPGAs

FPGAs are like pizzas you take home and bake yourself. With FPGAs, engineers do not customize anything on the chips physically. They simply use software to program the FPGAs to interconnect their elements as desired.

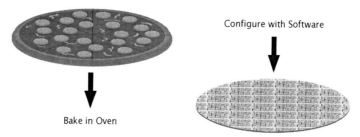

FPGA: No physical customization

FPGA is an acronym for field programmable gate array, which means:

- **Field.** Customized in the field, meaning anywhere outside of the IC fab.
- **Programmable.** Customized by software rather than by masks.
- **Gate.** A logic element made of transistors.
- **Array.** A large number of gates arranged in rows and columns.

The objective of developing FPGAs is to get projects done quickly and spend very little NRE. FPGAs satisfy these objectives because they do not need any customized masks at all. However, an FPGA's logic elements are very flexible and additional memory elements are needed to store configuration information, so the die size is much larger and the cost much higher than for an equivalent ASIC or full custom IC. FPGAs typically cost from one to five dollars. The largest FPGAs have hundreds of thousands of logic gates and cost several hundred dollars.

Selecting an IC class

The first table on the next page indicates how you can select the best class of IC for your project. The second table summarizes the characteristics of the classes that affect your selection. The main factor in choosing a class of IC is the quantity of parts you will need in mass production, because selling a larger volume of parts drives the need for a lower unit cost and provides profits to pay back a large NRE. High volume ICs need a low unit price but can tolerate a high NRE, so full custom ICs are the best fit. Low volume ICs can tolerate a high unit cost but need a low NRE, so FPGAs are the best choice. ASICs provide a balance of reasonable unit cost and moderate NRE, so they fit in the middle ground where production volumes are moderate.

Another important factor in choosing a class of IC is the time it takes to develop the chip. As shown in the second table, development time may be a few months for FPGAs, a year or so for ASICs, and several years for full custom ICs. Some programs take advantage of the best features of all three classes of ICs by prototyping with FPGAs, beginning low volume production with ASICs, and converting the design to full custom if the product is successful in the market and the volumes ramp up to millions.

Selecting an integrated circuit class

Annual production	Example application	Critical factors	Best IC class
1,000,000 or more	Cell phones	Low unit cost	Full custom
10,000 to 1,000,000	Medical equipment	Reasonable unit cost & NRE	ASIC
1 to 10,000	Specialized scientific instruments	Low NRE & quick time to develop	FPGA

Characteristics of integrated circuit classes

| Characteristic | IC Class | | |
	Full custom IC	ASIC	FPGA
Masks customized	All	A few	None
NRE[1]	$300,000 to $millions	$50,000 to $150,000	Less than $5,000
Die size[2]	Small	Medium	Large
Time to develop	2 to 5 years	About 1 year	A week to a few months
Unit cost, 10M/yr[3]	Small	Medium	Large
Unit cost, 100k/yr[3]	Not feasible[4]	Medium	Large
Unit cost, 1k/yr[3]	Not feasible[4]	Not feasible[4]	Large

Notes:
(1) Non-recurring engineering expense, the fixed cost of developing a chip.
(2) For equivalent functionality.
(3) IC cost per unit is shown for several levels of mass production, 10,000,000 per year, 100,000 per year, and 1,000 per year.
(4) IC manufacturers do not produce this type of IC at this low of a volume.

What does an ASIC engineer do?

If you have decided that ASICs are right for your project, you probably want to know what your role is in developing them. As an ASIC engineer, your job involves the following activities:

- Designing logic
- Using hardware description languages
- Turning HDL into silicon

Designing logic

The part of the job engineers enjoy the most is the area you probably studied in college, designing logic. In case you did not study it in college, this section introduces the basic concepts of designing logic.

Designing logic is the process of interconnecting logic elements so they function in a specified manner. Logic functions range in complexity from very simple, like deciding if either of two signals is true, to very complex, like executing Windows software. However, even the most complex logic function is built up from very simple ones, so you need to understand what simple logic functions are.

An example of a simple logic function is this egg sorting machine:

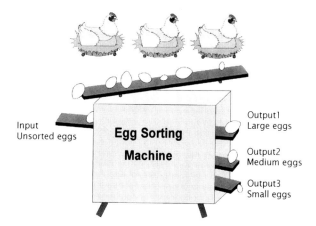

The egg sorting machine

The egg sorting machine accepts eggs of various sizes into its *input*, shown at the machine's upper left corner. (*Italics* indicate technical terms used by engineers who design logic.) The machine sorts the eggs and routes them to three separate *outputs*, shown on the right, one for large eggs, one for medium eggs, and one for small eggs.

You begin designing logic by drawing *block diagrams*, like the picture of the egg sorting machine. A *high-level* block diagram like this one uses a single block to represent the entire machine, and shows all its inputs and outputs. Engineers often call this block a *black box* (even though it is blue in this drawing), because you cannot look inside to see how it works. The block diagram is accompanied by a specification document that explains the functions, inputs, and outputs of the machine. The preceding paragraph is a specification of the egg sorting machine.

Once you understand the machine at a high level through your block diagram, your next step is *partitioning*. Partitioning means breaking down the functions inside the machine into smaller logic blocks, which you document in a more detailed, lower level block diagram and specification. (The egg sorting machine is so simple, it does not need to be partitioned.)

Partitioning also includes deciding whether to implement the entire machine in a single IC, or to put some blocks in separate chips or circuit boards instead. Engineers make decisions about partitioning by considering a number of factors including:

- **Cost.** How much the solution costs to manufacture in mass production.
- **Size.** How big the chips (or boards, or boxes) are.
- **Power.** How much power the solution consumes – how long it runs before the battery needs recharging.
- **Pin counts.** How many pins the chips need to carry all the input and output signals; or how many pins are on the connectors of boards and boxes.
- **Capabilities of IC manufacturers.** How many gates and how many pins IC manufactures can put on an chip.

For example, perhaps you can build your machine in a single chip, but if you split it into two smaller chips, the cost is lower. Based on cost, then, partitioning in two chips is the better solution. But your product may be a small cell phone that only has room for one chip; in that case, size is the most important factor and partitioning into a single chip is the best solution, even though the cost is higher.

After partitioning, the engineer's challenge is to configure logic elements to implement the functions required of each block. Looking inside the black box, you can see how the egg sorting machine works:

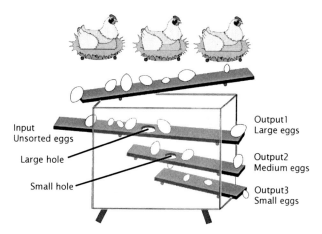

Input
Unsorted eggs

Large hole

Small hole

Output1
Large eggs

Output2
Medium eggs

Output3
Small eggs

Internal logic of the egg sorting machine

The egg sorting machine's hardware consists of boards with holes in them. The boards perform the following operations:

- **The top board.** Eggs roll past the large hole if they are larger than the large hole in the board. Thus eggs larger than the large hole roll to Output1, while eggs that are smaller than the large hole drop to the middle board.
- **The middle board.** Eggs roll past the small hole if they are larger than the small hole in the board. Eggs larger than the small hole roll to Output2, and since no large eggs reach the middle board, only medium eggs make it to Output2. Meanwhile, eggs that are smaller than the small hole drop to the bottom board.
- **The bottom board.** Eggs roll to Output3; since no large or medium eggs reach the bottom board, only small eggs make it to Output3.

A distinctive feature of the logic in the egg sorting machine is that each board's operation is a simple yes or no decision — is the egg larger than the hole, or isn't it? This is characteristic of *binary logic*, logic that uses only two states. The states of binary logic may be called True and False (as logic states), 1 and 0 (as mathematical quantities), or High and Low (as electrical voltage levels). Binary logic is the basis of all conventional logic design; it is the language that computers speak. They are called *digital* computers because they compute using the digits 1 and 0.

Another feature of the egg sorting machine is that it has two classes of behavior the engineer must analyze. *Functional* behavior is how the outputs change in response to changes in the inputs, irrespective of the time it takes to happen. The *timing* behavior is the amount of time it takes for an egg to get from the input to an output. This period is called the *propagation delay* through the system. If the egg sorting machine were driven by a clock, another important timing characteristic would be how fast the clock can run, like a Pentium processor that can operate with a 3 gigahertz clock but not faster. Engineers normally analyze functional and timing behavior separately, first getting the functional behavior to operate correctly, then working on timing issues.

As an engineer, your job is to design logic to implement particular functions like egg sorting, firing your engine's spark plugs at just the right time, or letting you play solitaire on your computer. Of course, you do not design ASICs using wooden boards with holes in them; you use logic *gates* which are made from transistors in ICs. Gates are very simple logic functions such as:

Logic gates: schematic symbols, equations, and functions

Name	Symbol	Equation	Function	Comment
AND	A B ⟶ C	C = A & B	The output is True only if *all* of the inputs are True.	
OR	A B ⟶ C	C = A | B	The output is True only if *any* of the inputs are True.	
Inverter	A ⟶ B	B = ~A	The output is True only if the single input is False.	Also called a NOT gate
NAND	A B ⟶ C	C = ~(A & B)	The output is False only if *all* of the inputs are True.	NAND means NOT AND
NOR	A B ⟶ C	C = ~(A | B)	The output is False only if *any* of the inputs is True.	NOR means NOT OR

You may think the egg sorting machine is a trivial example, but sorting and routing are fundamental operations in computing. Cisco Systems built a huge company sorting and routing data packets on the Internet!

Using hardware description languages

ASIC engineers specify logic with hardware description languages (HDLs), which are very much like software programming languages. Two HDLs, Verilog and VHDL, have been standardized and are used pervasively in the ASIC industry. The C software programming language is also used as an HDL to design ICs.

The egg sorting machine could be specified in an HDL as follows:

```
TopBoard: LargeEggs = Eggs > LargeHole;
MiddleBoard: MediumEggs = (Eggs < LargeHole) & (Eggs > SmallHole);
BottomBoard: SmallEggs = Eggs < SmallHole;
```

The elements in this HDL code are described in the following table:

Elements of sample HDL code

Element	Description
TopBoard MiddleBoard BottomBoard	Labels for the statements after the colons
Eggs	Symbolic name for the input
LargeEggs MediumEggs SmallEggs	Symbolic names for the outputs
LargeHole SmallHole	Symbolic names for the constants (fixed values) chosen to discriminate the sizes of the eggs
= > < &	Mathematical symbols for equals, greater than, less than, AND

The three statements in the sample HDL code read like plain English, or plain algebra. You can easily see they describe the operations performed by the boards and holes in the egg sorting machine.

Turning HDL into silicon

When you finish designing your logic, you still have a lot of work to do. You use these processes to turn your HDL description into an actual IC:

- Simulation
- Synthesis
- Layout

Simulation

You simulate your design by using a CAD tool called a *simulator* to examine the design's functional and timing behavior. You define input stimulus for your design as a set of *functional test vectors*; the simulator shows you how your design responds. Every time you change the design, you simulate again to verify the design still behaves correctly, or that improper behavior was corrected if that was the reason for the change.

The simulator can work with a *behavioral model*, which is an abstract representation of the design, or a *gate-level netlist*, which contains actual gates and timing information from the ASIC vendor's library of predefined logic elements. A gate-level netlist is produced by synthesis.

Synthesis

You use a CAD tool called a *design compiler* or *synthesizer* to create a version of the design in gates that are available on the ASIC base wafer. The design compiler accepts your HDL code and a library that describes the gates and *macros* (predefined functions made from gates) available in a your ASIC vendor's product. It produces a gate-level netlist which is a text file that lists the gates used to implement the design and the *nets*, or wires, that interconnect the gates. The design compiler reports the *gate count*, the number of gates used in the netlist. Minimizing gate count is important because a design with fewer gates can fit into a smaller ASIC. Also, a design with fewer gates is easier to lay out in an ASIC of a given size because it does not need to be packed as tightly as a design with more gates.

Layout

Your ASIC vendor uses CAD tools to turn the gate-level netlist into actual geometries for the chip masks that define the transistors on the IC. This process produces a more accurate timing model than the gate-level netlist previously contained, so you use the simulator once again to verify that the design's function and timing are correct before fabricating the chip.

Once layout is complete, your vendor can tool (physically create) the masks and fabricate the chips for you. When you receive the prototype chips, your HDL has been turned into silicon.

The following chart shows how the design is refined to different levels of abstraction as it moves from concept (top of chart) to implementation (bottom of chart):

Levels of abstraction as an IC is developed

Level	Example	Created by	Description
Specification	The Geneva cell phone is only 2 millimeters thick...	Engineers	Text description in plain English
Block diagram		Engineers	Partitioning the design into blocks that implement the specification
HDL	for (i=4'b0; (c = = 4'b1)) begin inc[i] = val[i] ^ carry; c = val[i] & carry; end	Engineers	Hardware description language describing detailed logic that implements the block diagram and specification
Gates		CAD tools	Logic gates synthesized to implement the logic described by the HDL code
Transistors		CAD tools	Transistors that implement the logic gates
Layout		CAD tools	Physical geometries that appear on the masks and are etched into the silicon surface to form and interconnect the transistors

How do I choose an ASIC vendor?

Your ASIC vendor is your partner as you turn your HDL code into silicon. You need to choose an ASIC vendor that can meet your needs and help you be successful. You should identify several ASIC vendors with product offerings that meet your project's requirements in these areas:

Factors in ASIC vendor selection

Factor	Question to ask
Gate count	How many gates do I need for my design?
IC package	What type of package and how many pins do I need?
Price	What does the vendor charge for the ASIC in mass production?
Performance	How fast does the chip need to run?
Power consumption	How much power does the chip consume?
NRE	What is the vendor's price to do the project?
CAD tools	Can the vendor work with my company's CAD tools? Or, does the vendor have a design center where I can use their CAD tools?
Development time	How quickly can the vendor do the project?
Business relationship	Are my company and the vendor willing to work together?

You begin your vendor screening process by looking at gate count and packaging. Vendors have various product families that may offer ASICs in sizes like 20,000, 50,000, 100,000, and 500,000 gates, and packages like 44-pin PLCC, 100-pin QFP, and 144-pin BGA (see pictures on page 8). Determine which vendors offer products with gate counts and packages that meet your needs; then look at the rest of the factors listed in the table. Try to identify several vendors who meet your requirements in all areas. The final decision usually comes down to picking the vendor who bids the lowest price for the ASIC in mass production.

You cannot choose your ASIC vendor on your own. Other people in your company need to be involved because a large amount of money is at stake, and because your company's business depends upon your vendor delivering product on time and at a good price. The final vendor decision is typically made by an executive in the materials (purchasing) department, taking into account your technical input.

You should select your ASIC vendor early in your development cycle if you can. Your vendor provides a library of logic elements that the design compiler tries to use efficiently as it synthesizes the gate-level netlist. However, you can help the design compiler do a better job if you are familiar with the library when you create your design. Therefore, select your ASIC vendor before you start designing if you can. If you do not know your ASIC vendor while you are designing the logic, you may need to go back and modify the design to achieve a lower gate count while you are synthesizing the gates.

What tools and equipment do I need?

Besides choosing a vendor for your ASIC project, you also need to choose your tools. You develop ASICs using expensive CAD tools running on powerful engineering workstations. The CAD tools needed by a single engineer may cost $50,000 to $100,000 or more. Most companies own or lease these tools if they do substantial ASIC development. Smaller companies or individual projects may use equipment at design centers owned by ASIC vendors.

The main CAD tools you need are a design compiler and a simulator. The design compiler, sometimes called a synthesizer, generates a behavioral model or a gate-level netlist from HDL code. The simulator allows you to examine the functional and timing behavior of your design. Besides these two workhorses, CAD vendors offer a wide variety of specialized tools to meet all kinds of challenges engineers face while developing ICs.

Your CAD tools run on engineering workstations like Sun SPARCs or PCs loaded up with a gigabyte or more of memory and ultra-fast disks and processors. These workstations usually run on the UNIX operating system, although Windows-based tools are becoming more powerful and popular. Make sure you have a big nineteen inch monitor so you can see a great deal of information on the screen at one time. If your project involves multiple engineers and workstations, the workstations need to be networked together and a fast, reliable server should be available to store your design data. Be sure you have a backup strategy to keep your valuable data safe in case of disk crashes, software viruses, or natural disasters.

A handful of vendors supply the CAD tools used by the majority of ASIC developers. Among the most popular CAD tool vendors are Cadence, Mentor Graphics, and Synopsys. All the vendors have Web sites where you can learn more about their products, and their salespeople are happy to visit you and explain why their tools are the best ones for doing your job.

Developing an ASIC

By now you understand what ASICs are and generally how they are developed. In this section you learn the specific process flow and procedures for developing an ASIC.

You develop an ASIC by following this process flow:

Start

| Designing the logic | → | Synthesizing the gates | → | Laying out the chip | → | Fabricating the prototypes |

Iterating the design

Verifying the prototypes

Performed by engineer

Performed by ASIC vendor

Problems found

No problems found

Done

You, as the engineer, perform the procedures of designing the logic, synthesizing the gates, and verifying that the prototype chips meets their requirements.

Your ASIC vendor does the actual job of laying out the physical geometries of the chip masks. You participate by verifying that their results correctly implement your design. Your ASIC vendor fabricates the prototype chips.

If the prototype chips fail verification, you need iterate the design, that is, you need to fix the design by repeating all the procedures. Iterating an ASIC is expensive and time-consuming, so make every effort to keep the number of iterations to a minimum. Most ASIC projects can be completed with only two or three revisions of the chip, including Rev. A, the first version.

Each of the procedures you must perform is broken down into a set of specific steps below.

Designing the logic

The objective of the design procedure is to produce a behavioral model of a design that implements the ASIC's functional requirements.

Before you can begin the design procedure, you need to develop a block level functional specification. In this specification, partition the design into small blocks. Each block should serve a single purpose and be simple enough to implement in no more than four pages of HDL code. Explain the function of each block in detail. Show all the inputs and outputs of all the logic blocks and how the blocks connect to each other.

After you generate the block level functional specification, you should approach the design procedure hierarchically. Treat each block as an individual design and perform the design procedure on it. When the blocks are working, combine them into higher level functions and perform the design procedure on those functions. Keep building higher level functions until you assemble all the blocks into a single chip and you can perform the design procedure on the whole ASIC. The hierarchical approach is effective because problems are easier to find and fix in smaller blocks than in a single large block.

```
12  parameter clk_cnt = 2;
13  function [7:0] increment;
14  input [7:0] a; reg [3:0] i; reg carry;
15    begin
16      increment = a; carry = 1'b1;
17      for (i = 4'b0; ((carry == 4'b1) && (i <=7)); i = i + 4'b1)
18        begin
19          inc[i] = a[i] ^ carry;
20          carry = a[i] & carry;
21        end
22    end
23  endfunction
```

A sample of Verilog HDL code
A complete ASIC has thousands of lines of code

Design the logic by performing these steps:

1 **Create the design.**

Write HDL code that describes logic which performs the specified function.

2 **If you are aware of any problems in the design, fix them.**

The first time you execute this step, you should not be aware of any problems. You return to this step if you find any problems in step 6.

3 **Compile the design using the design compiler.**

A behavioral model of the design is produced.

NOTE *Be sure to fix any errors the design compiler detects.*

4 **Do one of the following:**

- **If the functional test vectors do not exist yet, write them.**
  Functional test vectors are a set of input stimuli that thoroughly exercise the design.

- **If the functional test vectors already exist, adjust them as required for problems you found in step 6.**

5 **Simulate the behavior of the design using the simulator.**

The behavioral model responds to the test vectors with either correct or incorrect behavior.

6   **Examine the simulation results carefully to see if the design behaved in the manner you expect.**

Any unexpected behavior indicates problems that must be fixed.

7   **Repeat from step 2 until no problems are found in step 6.**

When you finish the design procedure, the design exists in an abstract logical form. It is complete and proven correct through behavioral simulation. The design is ready to be synthesized into specific elements available in your ASIC vendor's library.

Synthesizing the gates

The objectives of the synthesis procedure are to:

- Generate a gate-level netlist equivalent to the behavioral model you created in the design procedure.
- Minimize the gate count.

Synthesizing the gates means turning the abstract HDL description of your design into a gate-level netlist using elements from the library supplied by your ASIC vendor. Synthesis may be approached by working on individual functional blocks or by giving the design compiler the entire design at once. For some designs, you can achieve a lower gate count if you perform the synthesis procedure on individual blocks. For other designs, the design compiler does a better job working on the whole design. You should experiment with both approaches. You may achieve the best result if you synthesize some blocks separately and leave others to be synthesized with the entire design.

After you have synthesized the gates and you are satisfied with the gate count, you need to verify that the resulting gate-level netlist behaves correctly. Problems may arise because the design compiler interprets your HDL code differently than the simulator does, or simply because the design compiler fails to generate some functions the way you want. In this procedure, you repeat simulating, fixing the design, and synthesizing the gates again until you achieve a result that behaves correctly.

NOTE *You must select your ASIC vendor before performing the synthesis procedure. Your vendor supplies the logic element library you need for this procedure.*

Synthesize the gates by performing these steps:

1 **If you are aware of any problems in the design, fix them.**

The first time you execute this step, you should not be aware of any problems. You return to this step if you find gate count problems in step 2, functional problems in step 5, or timing problems in step 8.

2 **Synthesize the gate-level netlist from the design's HDL description using the design compiler.**

A gate-level netlist is produced and the design compiler reports the gate count.

3 **Repeat from step 1 until you are satisfied with the gate count in step 2.**

NOTE *Iterating synthesis several times while changing the design in ways that may help reduce the gate count is usually a good idea.*

4 **Simulate the design at a functional level using the simulator.**

The gate-level netlist responds to the test vectors you prepared in the design procedure with either correct or incorrect behavior. Simulate with relaxed timing (short gate propagation delays and slow simulated clock speeds) so you can focus on functional problems, not timing problems.

5 **Examine the simulation results carefully to see if the design behaved in the manner you expect.**

Any unexpected behavior indicates problems that must be fixed.

6 **Repeat from step 1 until no problems are found in step 5.**

7 **Verify the timing using the simulator.**

The gate-level netlist responds to the test vectors you prepared in the design procedure with either correct or incorrect timing. Simulate with worst-case timing (long propagation delays and fast clocks) so you can detect timing problems.

8 **Examine the simulation results carefully to see if the design responded with the timing you expect.**

Any unexpected timing indicates problems that must be fixed.

9 **Repeat from step 1 until no problems are found in step 8.**

When you finish the synthesis procedure, the design exists in a form that accurately models the characteristics of your ASIC vendor's library elements. However, it does not accurately reflect the timing of the final chip because the CAD software had to estimate the lengths of the interconnect that will actually hook up the gates on the chip when it is laid out. (See sidebar, *Interconnect and timing*, on the next page.) To model the timing accurately, the physical geometries of the chip must be known. Hence, the next step is laying out the chip.

Interconnect and timing

Interconnect is wiring that hooks up gates on a chip, and it makes a big difference in how fast signals change. At the microscopic scale of an IC, signals do not change abruptly like flipping a light switch on and off. Rather, they change more slowly, like turning a dimmer switch gradually from off to dim, brighter, and full on — and an IC's tiny transistors cannot move the dimmer switches very quickly.

The more interconnect a transistor has to drive, the slower the dimmer switch turns, slowing the speed signals travel through the chip. Engineers say the interconnect is *loading* down the transistor, or the interconnect is a *load* on the transistor.

Because loading affects signal speeds so heavily, interconnect must be modelled precisely to accurately simulate a chip's timing. But the interconnect configuration is not known until the chip has been laid out. How can you examine the timing before laying out the chip?

ASIC vendors solve this dilemma by statistically analyzing the interconnect loads of many completed chips. For example, if a particular type of gate has an interconnect length of less than 10 microns in 90 percent of the cases they examined, you can assume your design is likely to be the same. So the CAD tools use 10 microns as a conservative estimate for all the gates of that type in your design. These estimated characteristics are used for pre-layout timing analysis.

Because these estimates of interconnect loading come from empirical, statistical data, pre-layout timing analysis is said to use *empirical timing* or *statistical timing*.

Laying out the chip

The objective of the layout procedure is to generate the actual geometries that form the masks to build the chip.

In the design and synthesis procedures, you decided exactly what toppings you want on your ASIC pizza. Laying out the chip is like putting the toppings on the pizza. Your ASIC vendor uses powerful CAD tools to determine how to actually arrange the gates on the ASIC die and interconnect them according to your gate-level netlist. The layout process defines the actual patterns that are created on the chip. When this task is completed, CAD software extracts the exact interconnect loads for all the gates on the chip. You use the design compiler to *back annotate* these values into your gate-level netlist, that is, to add that information into your design. Then you simulate the function and timing again to verify the design still behaves correctly.

NOTE *Your ASIC vendor must lay out the chip before you can start this procedure. You need to give them a purchase order for 25 percent of the total NRE charges before they start. (See sidebar, NRE payment schedule, on page 32.)*

Participate in the layout process by following these steps:

1 **If you are aware of any problems in the design, have your vendor fix them.**
   The first time you execute this step, you should not be aware of any problems. You return to this step if you find functional problems in step 4 or timing problems in step 7. You must determine how to fix the problems; your vendor implements the fixes you specify.

NOTE *During the layout procedure, your vendor can fix problems by optimizing certain parts of the layout, using larger transistors to speed up signals, or inserting gates to slow down signals. If you encounter problems which cannot be fixed with these techniques, you must abort the layout procedure and go back to the beginning of the synthesis procedure.*

2 **Generate the post-layout netlist.**
   Use the design compiler to back annotate the interconnect loading information provided by your vendor into your gate-level netlist. If any fixes were made in step 1, your vendor gives you a new set of values to back annotate.

3 **Simulate the design at a functional level using the simulator.**
   The post-layout netlist responds to the test vectors you prepared in the design procedure with either correct or incorrect behavior. Simulate with relaxed timing (short gate propagation delays and slow simulated clock speeds) so you can focus on functional problems, not timing problems.

4   **Examine the simulation results carefully to see if the design behaved in the manner you expect.**

Any unexpected behavior indicates problems that must be fixed.

5   **Repeat from step 1 until no problems are found in step 4.**

6   **Verify the timing.**

The post-layout netlist responds to the test vectors you prepared in the design procedure with either correct or incorrect timing. Simulate with worst-case timing (long gate propagation delays and fast clocks) so you can detect timing problems.

7   **Examine the simulation results carefully to see if the design responded with the timing you expect.**

Any unexpected timing indicates problems that must be fixed.

8   **Repeat from step 1 until no problems are found in step 7.**

NOTE   *Your ASIC vendor may generate the post-layout netlist (step 2), simulate the design (steps 3 & 4), and verify the timing (step 6 & 7) for you, and simply inform you of any failures they find. This saves time because of the logistics involved in moving the design data back and forth between you and your vendor.*

When you finish the layout procedure, the design exists in its final form as it is intended to be built on the actual chip, and it has been verified to work correctly in both its functional and timing characteristics. You are ready to sign off the design and send it to fab. Have a tapeout party! (See sidebar, *Tapeout*, on the next page.)

Signing off the design

The objective of the sign-off process is to approve the design and release it to your ASIC vendor to authorize them to fabricate prototype ICs for you.

Signing off the design consists of preparing the documentation listed in the table below. When the chip has been signed off and the sign-off documentation has been transmitted to your ASIC vendor, they fabricate prototypes of your chip.

Sign-off Documentation

Document	Originates from	Description
Final netlist	Design compiler	Defines the final configuration of the gates
Functional test vectors	Design procedure	Used for testing the chip
Timing checks	Timing verification in layout procedure	Used to establish timing goals for testing the chip
Package marking specification	Engineer	Artwork for the company logo, part number, and other text you want printed on the IC package
Authorization forms	ASIC vendor	Various forms your vendor asks you to fill out
Purchase order	Your company	Pay 25 percent of the NRE charges (see sidebar, *NRE payment schedule*, on page 32)

Tapeout

When IC designers release a chip for fabrication, they say they are *taping out* the chip, or they have reached *tapeout*. This term comes from the time before the internet. The chip data would be stored on a big reel of magnetic tape, and the tape would be sent to the fab. The tape literally went out the door!

A 9-track, 10.5" diameter, open reel tape

Fabricating the prototypes

The objective of the fabrication process is to manufacture prototype ICs.

Vendors run prototypes on special lines for speed and quality control, and so ongoing mass production is not disrupted. They run two or three wafers of your chip so if something goes awry on one wafer, the other wafers may still yield good parts. ASIC vendors include twenty-five samples of your chip in your NRE charge, but you can negotiate for more if your program demands it.

Turn around time for prototypes is four to eight weeks. Two weeks of that time is required just to package the completed die. You need to give your vendor a purchase order the remaining 50 percent of the NRE when they deliver the prototypes. (See sidebar, *NRE payment schedule,* below.)

Once the prototype chips have been fabricated and delivered to you, your next task is to verify that they work.

NRE payment schedule

ASIC vendors require you to pay the NRE charges in several installments as the program progresses. 25 percent of the NRE is due when they begin laying out your chip. 25 percent more is due when the begin fabricating your prototypes. The remaining 50 percent is due when they deliver the prototype chips to you.

These payments are non-refundable. If you cancel the program, you forfeit any payments you have made, because these payments compensate the vendor for the work they completed on your project.

The timing of the NRE payments is shown below:

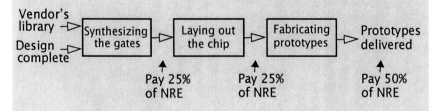

Verifying the prototypes

The objective of the verification process is to prove that the prototype ICs meet all their functional, timing, and other (environmental, mechanical, etc.) requirements, and that they meet the needs of the system they were designed to be part of.

When your receive your prototype chips, you need to evaluate them quickly. Many people are anxious to hear the verdict because a large amount of time and money was invested, and because the program will be delayed if the part has problems. In order to verify the chip rapidly and efficiently, you should have a test system known as a *test bed* ready when the prototypes arrive. The test bed should exercise the chip the same way as it is exercised in the final product. Sometimes the final product itself can be used as the test bed.

You need to verify that the chip behaves as designed, and also that it operates correctly in the system for which it was designed. Exercise the chip in all the conditions it is likely to encounter in all of its targeted uses. Vary external conditions such as voltage and temperature to ascertain the chip's quality and design margins (how far the specifications can be extended without failures).

Verification is an important and complex task. Formulate your test strategy early in the development process; you may want to design special features into the chip to help you verify it. You may build the test bed hardware and write any software you need during the weeks while your prototypes are being fabricated. However, verification may be such a big job that preparations must be started much earlier. Other hardware and software engineers may even be assigned to this task.

If verification testing proves the chip meets all of its specifications and it functions correctly in its target environment, the process of developing the ASIC is complete. Release the design to manufacturing so they can order parts for production. However, problems often show up during verification. Sometimes they can be tolerated, patched, or worked around in software; if not, you have to iterate the design to fix them.

Iterating the design

The objective of iteration is to revise the design until you obtain prototype chips that pass verification.

CAD tools and ASIC vendors are so good these days, first silicon (the first set of prototypes) often yields good parts. However, when bugs do show up, you need to eliminate them by revising the design and prototyping the chip again. This process is known as iterating the chip, or stepping the chip. What it means is that you return to the beginning of the design procedure and do it all over again. Subsequent iterations can be accomplished much more quickly than the first pass because you can focus on just the few problems you are fixing. Nevertheless, each step of the development process must be repeated in order for you to be successful. Shortcutting the process is a recipe for disaster!

Given the complexity you are dealing with when you develop an ASIC, at least one bug has a substantial probability of showing up — no matter how well you do your job. Therefore, you should always plan your program to allow you to revise your chip one time. Another revision beyond even the one you planned may be necessary, but if you do your job well, you should never have to revise your chip more than twice. In any case, your first silicon should be good enough to support most of the system testing required for your program.

Moving On

The Appetizer section (*Getting Started*) reviewed what ASICs are, how they are manufactured, and what the engineer's role is in developing them. The Entrée section (*Developing an ASIC*) explained the process flow and procedures an engineer follows to develop an ASIC. However, you need to know much more before you actually start developing your first ASIC. So, for dessert, here are some suggestions of how you can learn more about developing ASICs.

Hardware definition languages

Excellent books on the Verilog and VHDL hardware description languages can be found at Amazon.com and other book stores. Some books come with CD-ROMs containing software you can use to simulate your HDL code and see how it actually behaves. Classes are also available from a variety of sources.

CAD tool vendors

Visit the Cadence, Mentor Graphics, and Synopsys Web sites to learn about the tools you use to develop ASICs. Plan to take training classes from your CAD vendor so you can get the most benefit from these sophisticated tools.

ASIC vendors

Visit the Web sites of vendors like VLSI Technology and Toshiba to learn about their ASIC products and training classes. Study the data books of their ASIC products; these books are essential tools you need to be intimately familiar with when you develop those ASICs. Invite some ASIC vendors to visit you to explain their product offerings and drop off their data books.

FPGA vendors

Visit the Web sites of programmable logic vendors like Xilinx and Altera for information about designing logic for FPGAs; most of it applies to designing ASICs as well. You may want to experiment with FPGA development tools. Getting set up for FPGA development is much less expensive than for ASICs. You may be able to obtain or borrow tools for free. You may even want to model some of your logic in FPGAs before completing your ASIC design.

Company resources

Think about training opportunities within your company, such as:

- Attending design reviews of other ASIC projects.
- Borrowing ASIC vendor or CAD tool data books or training manuals from other engineers.
- Studying the HDL code and simulation test benches from other ASIC projects – or full custom IC or FPGA projects.
- Offering to review HDL code or help with simulations on another project, on your own initiative or as a temporary work assignment.

The best way for you to become an excellent ASIC engineer is through real world experience. However, if you are an independent consultant or the lone ASIC engineer in a small company, you may not have access to these types of company resources. Do not despair. The CAD tools and the support from the ASIC and CAD vendors are so good that even inexperienced engineers can be successful on their very first ASIC designs if they approach their projects with intelligence, engineering discipline, and great attention to detail.

Cleanup

Troubleshooting

Just as cooking a three course meal or a pizza leaves a mess
to clean up, you are bound to run into some messes when you
develop ASICs. The table on the following page lists some
common problems you are likely to encounter as you develop
ASICs, and recommends solutions for them.

Troubleshooting

Problem	Description	Solution
Failure to meet timing goals	The logic you design may fail to operate at the highest clock speeds or with the slowest propagation delays required by your product specification. Meeting your timing goals is referred to as achieving *timing closure*.	Change the design and rerun synthesis and layout until timing closure is achieved. Get help from other engineers. If absolutely necessary, change the chip specifications to relax the timing requirements or find a faster ASIC technology.
Bugs in the CAD tools	CAD tools are extremely complex programs so they inevitably have bugs (problems) which affect your ability to do your job.	Work with your CAD tool vendors to fix, patch, or work around the problems.
High cost of CAD tools	The CAD tools used by a single engineer may cost $50,000 to $100,000 or more. A project's return on investment (ROI) may not justify such a large expenditure.	Spread out the cost burden by using *floating licenses* (see sidebar, *Floating licenses*, on next page) to share the software among engineers working on multiple workstations and projects.
Confusion over CAD tool licenses	You cannot access the floating licenses for resources you need (see sidebar, *Floating licenses*, on the next page).	Be careful to request the correct licenses and to follow the exact procedures your IT people prescribe. Work with your IT people to resolve problems.
Poor interoperability between CAD tools	CAD tools from different vendors may not be able to exchange design data because they: • Use different data and file formats. • Interpret parts of standardized formats differently. • Support different feature sets. For example, a simulator from Vendor A cannot load the netlist generated by a design compiler from Vendor B.	Ask in-house experts to write scripts (small programs) to patch or work around the problems; or write such scripts yourself. Work with your CAD tool vendors to solve these issues.

Floating licenses

CAD tool software can be shared by engineers working on multiple workstations by using a mechanism called *floating licenses*. Floating licenses are software licenses that are not tied to a particular user or workstation. They take advantage of the network connecting the workstations and servers to "float" around to where they are needed at any particular time.

Engineers check out floating licenses from the network when they need to use licensed resources. They check the licenses back in to release the resources when they done using them.

With licenses for different software and different versions of software floating around the network, confusion often arises. As a result, engineers may be unable to access critical resources.

You can avoid most licensing confusion if you carefully identify which licenses you need, and you meticulously follow the procedures your IT department has set up for checking licenses in and out. Also, be a good citizen and release your licenses promptly, so others can use the resources you no longer need. When you do have problems with licenses, ask your IT people for help.

Glossary

Application specific integrated circuit (ASIC)

An integrated circuit that can be developed quickly and at a low cost; an ASIC usually serves a single, specialized purpose.

Back annotate

To modify a design model to include a new set of characteristics; for example, to back annotate the netlist with information about interconnect loading extracted from the chip layout.

BGA

Ball grid array; a type of IC package that packs a large number of pins in a small area; the pins are actually tiny balls of solder.

Binary logic

Logic that works on a set of only two values called True and False, 1 and 0, or High and Low; *digital* computers operate using binary logic.

Black box

An abstract representation of a logic function which specifies the inputs, outputs, and logical transformation, but not the internal structure that implements the transformation.

Block diagram

A drawing that indicates the inputs, outputs, and interconnection of *black box* logic functions.

Bug

A problem in electronic hardware or computer software; term comes from an actual dead moth that was discovered in a failed computer in the early days of computing; also can refer to an IC because the little black packages with leads sticking out on two sides resemble insects, like this DIP package:

CAD

Computer aided design; software programs used as tools by engineers developing ASICs, and in other engineering disciplines as well.

Chip

Informal term for an IC or a die; term comes from the tiny piece (like a chip of stone) of silicon that has a large number of electronic circuits integrated on it.

Clock

A constant, repeating signal, like the tick of a timepiece, that drives computer logic at a fixed speed; for example, a 3.3 gigahertz Pentium has a clock that runs at 3.3 gigahertz (a gigahertz is 1,000,000,000 or 10^9 cycles per second).

Design

The intellectual process of creating logic, circuits, or other implementations to meet a set of specifications; also can refer to the end result of the process of designing; sometimes the entire development process is referred to as design.

Design compiler

A CAD tool that generates logic gates which implement logic described in an HDL; also called a *synthesizer*.

Die

A rectangular silicon chip; an IC without its package. Die is used for both singular ("put the die in the package") and plural ("this wafer has 200 die on it") forms.

Digital

Working in a domain of discrete values, for example, binary logic.

DIP

Dual in-line package; a type of IC package that has two rows of pins; called a *through hole* technology because the pins stick through holes in the circuit board. (see illustration under *Bug*).

Empirical timing

See *statistical timing*.

Fab

Fabrication facility; a manufacturing plant for building ICs; inside a fab are *fab lines* which are production lines for processing silicon wafers into ICs; also, as a verb, shorthand for *fabricate*, as in "to fab an IC."

First silicon

The first set of prototype chips received when a new design is fabricated.

Functional behavior

The operation of a logic machine independent of its *timing*.

Gate

Very simple logic functions like AND, OR, NOT, NAND, and NOR, which are defined as follows:

- The output of an AND gate is True only if *all* of the gate's inputs are True.
- The output of an OR gate is True only if *any* of its inputs is True.
- The output of a NOT gate (also called an *inverter*) is True only if its single input is False.
- The output of a NAND gate is False only if *all* of its inputs are True; NAND means NOT AND.
- The output of a NOR gate is False only if *any* of its inputs is True; NOR means NOT OR.

Gate count

The number of gates on a chip, normalized to a 2-input NAND gate.

A 2-input NAND gate is implemented with four transistors and is the fundamental unit used for counting gates on an ASIC. Here are a few examples:

- An inverter counts as 0.5 gates because it is made with two transistors.
- An AND gate counts as 1.5 gates because it is made of a NAND gate followed by an inverter.
- A 4-bit counter *macro* (it counts from zero to fifteen) may have a gate count of 80 gates.

HDL

Hardware description language; an English-like language for specifying logic; similar to a software programming language.

High-level block diagram

A block diagram that shows only the overall inputs and outputs of a function; it has not been *partitioned* into smaller sub-blocks.

Hydraulic knife

A thin, very high pressure stream of water used to cut apart the die on a silicon wafer.

Input

A signal that goes into a logic function or gate.

Integrated circuit (IC)

An electronic component which has a large number of devices like transistors built onto a tiny piece of silicon; also called a *chip*; IC can refer to just the chip of silicon or to the chip in a package like a *BGA, DIP, PLCC,* or *QFP.*

Interconnect

Refers to the lines of metal (or other material) that connect the devices on an IC; essentially wires.

K

Abbreviation for 1,000 or 1,024 (2 raised to the 10th power).

Layout

The physical geometries of the devices on a chip, as defined by the chip's masks; *laying out* the chip means designing these geometries.

Library

A computer data file containing descriptions of the *gates* and *macros* available on an ASIC; the descriptions include both functional and timing characteristics of the elements.

Load

The amount of interconnect a transistor or a gate has to drive; long interconnect wires are big loads that slow down how quickly transistors or gates can change their output states.

M

Abbreviation for one million or 1,048,576 (2 raised to the 20th power).

Macro

A pre-designed, complex logic function like a multiplexer, counter, register, or memory; ASIC vendors' libraries have rich sets of macros for engineers to design with.

Margins

Safety factors built into designs; for example, if a design is specified to operate over the voltage range 4.5 volts to 5.5 volts, but testing proves that it works over 4.0 volts to 6.0 volts, the design margin for operating voltage is 0.5 volts.

Mask

An optical stencil used to define the fine geometries of transistors and interconnect on an IC.

Netlist

A text file listing how a set of devices are interconnected to form a particular electronic circuit; a *net* is all the points a particular signal is connected to.

NRE

Non-recurring engineering expense; the amount invested in developing an IC; in accounting terms, NRE is a fixed cost because it does not vary with the number of units manufactured.

Output

A signal that comes out of, or is generated by, a logic function or gate.

Partitioning

The process of breaking down a function into smaller sub-blocks and deciding where to implement each block; for example, a desktop computer's electronics are partitioned into a motherboard and several add-on cards.

Patch

To fix a problem temporarily, in a "quick and dirty" fashion.

PLCC

Plastic leadless chip carrier; a type of IC package with pins that do not stick through the circuit board (it is a *surface mount* component, i.e., mounted on the surface of the circuit board).

Propagation delay

The time it takes for the output of a gate or a series of gates to change after an input changes; for example, a particular AND gate may have a propagation delay of 5 nanoseconds (five thousandth of a microsecond or 5×10^{-9} seconds) when it is driving a specified amount of *interconnect*.

QFP

Quad flat pack, a type of IC package which can have over 300 pins.

Schematic

An abstract drawing that shows how electrical components like transistors or logic gates are interconnected, using symbols like this:

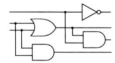

Script

A small program written in a scripting language like Perl, often used to pre- or post-process data moving between CAD tools; scripts have the advantage of being "quick and dirty" compared to other types of software.

Semiconductor

A characteristic of silicon crystals containing certain impurities; such crystals conduct electricity only when a particular electric field is applied, and they can restrict current to flow in only one direction; generally refers to ICs made from such crystals.

Silicon

A chemical element that forms crystals which can be processed into semiconductor devices.

Simulator

A CAD tool that shows how a logic design responds to a set of input stimulus or *test vectors*; a *behavioral* simulator works with a model of the design expressed in an HDL; a *functional* simulator works with a model expressed in gates; a *timing* simulator models the speeds of signals as well as their functional behavior.

Statistical timing

Timing estimates used for pre-layout timing analysis; the estimates are based on statistical analysis of the interconnect in previous designs; also called *empirical timing*.

Synthesizer

See *design compiler*.

Tapeout

Releasing design data so prototype ICs can be fabricated; at one time the data would be transferred on magnetic tape and the tape was literally sent out to the fab.

Test bed

A lab fixture for testing prototype ICs; sometimes the product for which the IC was designed can serve as the test bed.

Test vectors

Input stimulus patterns for a logic simulator; may also list the expected (correct) states of the outputs; a *vector* is a one dimensional array, so a single test vector is a string of 1s and 0s that defines for a particular moment in time the state of all of the inputs and outputs of the device under test.

Timing

The speed at which signals change and propagate in an IC.

Timing closure

Satisfying the timing requirements of the design; timing closure is achieved when the design operates correctly at the highest clock speeds and with the slowest propagation delays required by the chip's specifications.

Tooling

Creating a tool used to produce a unique part; like creating the molds for a plastic part or the masks for a semiconductor IC.

Transistor

A three terminal semiconductor device which can be used as a switch or an amplifier.

UNIX

The most popular operating system for engineering workstations and network servers; an alternative to Windows, although Windows is becoming more powerful and popular in the engineering community.

Verilog

A popular HDL used to design ICs.

VHDL

VHSIC (very high speed integrated circuit) hardware description language; a popular HDL used to design ICs.

Wafer

A thin, polished disc of crystallized silicon which is the starting material for manufacturing ICs; state-of-the-art fabs process wafers which are 300 millimeters or 12 inches in diameter; the cover of this guide has a picture of a silicon wafer with ICs fabricated on its surface.

Work around

A procedure to temporarily deal with a problem without fixing the root cause.

Index

About the Author

Elaine Rhodes developed and
used integrated circuits and
ASICs for twenty-five years in
Silicon Valley high-tech companies like Intel, Tandem
Computers, Quickturn Design Systems, Exabyte, and
NeoMagic. With the publication of ASIC BASICS, she
begins a second career as a technical writer. Ms. Rhodes
lives in San Jose, California, with her calico cat,
Calpurrnia. She invites you to visit her personal Web
site, www.elainerose.com, or write to her at
lannirose@gmail.com.

Lightning Source UK Ltd.
Milton Keynes UK
UKOW01f1915190217
294769UK00001B/313/P